The little animal pictured on the next page is a rat-tailed dwarf lemur. You are probably already familiar with some kinds of lemurs from books, movies and zoos. I am a biologist, and I've spent many years studying lots of different lemurs.

For relaxation on weekends, my friend Andrew and I often went for long bike rides. We would entertain one another while riding by talking about the work we had done in the preceding week. Andrew directed a clinic for the study of sleep disorders. Everything I know about sleep came from Andrew.

Sleep is necessary for survival. We're not sure why, said Andrew, but if we humans are denied sleep for any length of time, we become mentally confused. Then we lose the ability to maintain our body temperature, and death comes soon after. In fact, this is true for all warm-blooded animals. Birds and mammals, though not reptiles, fish or frogs, must sleep in order to survive.

A fat-tailed dwarf lemur (*Cheirogaleus medius*), held by a researcher at one of our field sites in Madagascar. This little animal has been the subject of our studies for several years.

Sleep is an active process, and like all activity, requires energy. Scientists can tell when a subject is sleeping because a particular pattern of brain activity occurs, which can be measured with small electrodes placed on the head and displayed on a computer screen. This pattern is known as the Delta rhythm. It is accompanied by another rhythm that comes in short bursts, and is known as REM, or rapid eye movement. In humans, REM is associated with dreaming.

You've probably heard of hibernation. Hibernation literally means "winter sleep". It is a state of lowered activity of bodily organs, so the body requires less energy to survive. Hibernating animals may appear to be sleeping, but they usually are not, and here is the big mystery. During sleep, the brain is still active. This means that it requires energy, either from food, or from stored fat. When a warm-blooded animal which is hibernating gets too cold, nerves cannot function. Without them, sleeping is not possible. But they don't die! This is interesting, I thought! What about animals such as the Arctic ground squirrel, that hibernate at very low temperatures? I asked Andrew "do these animals manage to get along without sleep during their hibernation?" Andrew did not know about Arctic ground squirrels.

"Perhaps," he said, "they are an exception. Anyway, I know sleep is essential for primates."

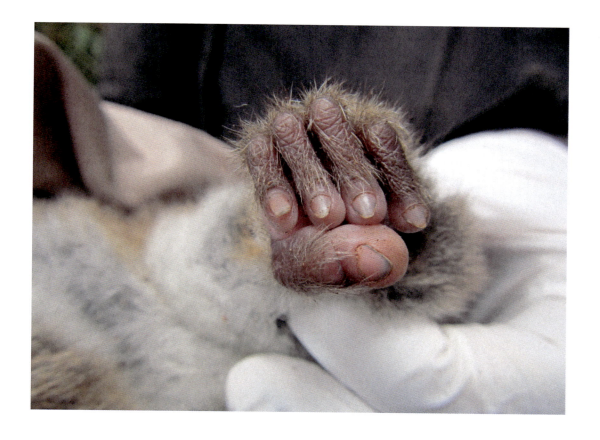

The hand of a fat-tailed dwarf lemur. Note the similarity to a human hand-- fingernails and an opposable thumb!

But lemurs are primates too, and the fat-tailed dwarf lemur definitely hibernates. A colleague of mine had confirmed this, just that week, at a research site in Madagascar where the temperature was too cold to allow nerves to function.

How come they survived without sleeping?

A hibernating bear lowers its body temperature just a few degrees; enough to be able to sleep (and thus stay alive), while needing less energy than it would if it was awake. This way it can live off its fat stores for the winter without needing to eat. Hibernating lemurs, however, like the Arctic ground squirrels, become fully torpid: that is, their body temperatures are identical to the temperature of their surroundings. They become as cold-blooded as a snake. They may become so cold that their need for energy drops to only 1-2% of their normal requirements. Their heart rate falls from over 200 beats per minute to just 6-10 beats. If you picked one up, you might not be sure it was still alive!

This Sifaka, one of the larger lemurs, has been torpid and is now being helped to warm up

So then, how can they possibly sleep? Yet without sleep, they should die.

Andrew had no answer to this mystery. My immediate thought was, let's explore this! I'd always wanted to go to Madagascar, the only place in the world where lemurs occur naturally. Maybe we could figure out how these little animals were surviving hibernation.

All lemurs evolved in Madagascar, a large island that was formed when a piece of continent broke off about 60 million years ago and drifted to its present position off the coast of eastern Africa. The ancestors of today's lemurs probably floated across the ocean onto the island sometime later. From them, all present-day lemurs are descended. That island is their only natural habitat. And it is incredibly varied, ranging from thick wet rain forest to cool mountain highlands to strange arid deserts. Some regions get almost no rainfall, while in others, it never seems to dry out.

The island of Madagascar in its present position east of Africa

Central highlands (top) and eastern rainforest of Madagascar

Tsingy de Bemaraha National Park in the northwest of Madagascar

Lemurs come in all shapes and sizes. The largest, now extinct, was almost as large as a human; the smallest is the size of a house mouse! There are over 100 known species, and more are discovered every year. Each species is well adapted to the varying conditions across the country.

The fat-tailed dwarf lemur lives in several different habitats. Some reside in the dry forests in western Madagascar, an area where food (fruit and leaves) is only available during part of the year. Others live in the highlands, where winters are long and cold. A third group lives in the rain forests, where their food supply is also seasonal. This means that for up to half a year, there is nothing for the lemurs to eat. How do they survive? They hibernate! They crawl into a hole in a tree, or in the ground, and go into a state of torpor where they don't need to consume energy. Their bodies become as cold as the air around them.

Ring tailed (above) and brown lemurs

A Sifaka, one of the larger species

Andrew, you will recall, had insisted that animals that were deprived of sleep could not survive. But a hibernating lemur spends up to half a year in that state. That is an impossibly long time to go without sleep. How do they do it? Andrew agreed that we should try to find out, especially if it gave us an excuse to go to Madagascar!

A mouse lemur, one of the smallest species

Before going to Madagascar, we needed to learn how to make the necessary measurements. Fortunately, the Duke Lemur Center had a small number of fat-tailed dwarf lemurs, the only colony we knew of anywhere in the world. We began by finding ways to stimulate them to hibernate in the climate of North Carolina. Shortening the days did not do it, and lowering the temperature didn't either. Nor did doing both together. How about food shortage, as commonly occurs in with the onset of winter? Federal regulations governing the care and use of animals such as lemurs didn't allow us to withhold food. Fortunately, one of the members of our research team, recognizing that the food source in the wild changed with the seasons, suggested we make a similar change in what had been a standard year around diet. Sure enough, adding more sweet fruits did the trick! With the change in diet the shortened days did induce the animals to hibernate. We think diet has its effect by changing the bacteria in the gut. This microflora does change in nature, though we cannot yet be certain it is the cause of hibernation.

A fat-tailed dwarf lemur with Dr. C. Williams and PHK

This fat-tailed lemur is ready for hibernation, having filled its fat stores in readiness. After a few months its tail will be slender again.

Next, we had to discover a way to record brain waves, to determine whether and when the animals were sleeping. We tried using the electrodes doctors employ for recording brain waves in human babies, but fat tailed lemurs are about the size of squirrels, and there simply wasn't enough room on their heads for the minimum number of electrodes needed. Finally, after much discussion, the authorities who regulate the use of animals for research decided that it would be safe and humane to use super fine needles, placed just under the scalp. This worked just fine, and we were able to record brain waves cleanly. The same electrodes also allowed us to record heart rate, which gave a hint as to how actively the animal's systems were functioning. To back up this latter information we also purchased a box large enough for the lemur that allowed us to record the amount of oxygen it was consuming. Finally, we fitted each animal with a comfortable collar that transmitted the body temperature to our recorders.

The chamber we designed for our hibernating lemurs. Air is pumped through and then analyzed so we can measure how much they are breathing

With our techniques now refined, we packed our gear and took off for Madagascar. We landed in the capital, the crowded city of Antanarivo, then traveled south and west to the coastal town of Morandava. From there, it was an arduous days' drive north to a former forestry station, now a research center called Kirindi, run by a German University. That became our home for several weeks.

A hilltop view of Antanarivo

Baobab Alley, the road to Kirindi (top). And an uninvited dinner guest (a fossa) at our research station in Kirindi

Kirindi has a temperate climate. It's warm and wet in the summer months and cool and dry in the winter. In the winter, June through September, it does get cold, but not quite down to freezing.

Thus, we decided we needed to repeat our study in more extreme environments. After two seasons at Kirindi, we moved our field camp to Tsinjoarivo, in the Central Highlands. This area has an elevation of about 2000 meters and has forests that harbor many fat-tailed lemurs. It is the highest part of Madagascar and the temperatures in the winter are often at freezing. It's also usually cold and wet during the day, just the sort of climate where even we would like to hibernate!

We were also seeking fat-tailed lemurs in a tropical rainforest climate. We found such an area in Marojejy, on the east coast of Madagascar. There we established our third and final field station.

PHK and assistant trying to stay warm in Tsinjoarivo

Our research site in Marojejy

You will remember, our question was: do lemurs sleep while hibernating? As expected, our recordings showed that as long as the temperature of the environment, as well as their body temperature, was above about 25°C. they did show normal sleep rhythms. However, once the temperatures dropped significantly lower, brain activity slowed or stopped. To our surprise, every few days the animals would appear to come out of hibernation, though not fully. For about 6 to 12 hours every 3 to 10 days depending on how cold it was, they would raise their body temperature to about 30 or so degrees C (38C is near normal), but only for a few hours! During this time the brain would display intense bouts of sleep. Especially interesting, the measurements showed the REM pattern; the pattern that, in humans, indicates that dreaming is taking place.

Of course, we have no way of knowing whether lemurs dream. But these bouts of increased metabolism use a lot of energy. In fact, most of the energy that the animals require to survive 5-7 months of hibernation is needed for these periods. Evolution generally eliminates wasteful efforts. It is hard to imagine such bouts would persist if they were not necessary. They seem to underscore the importance of sleep, even for a hibernating lemur.

There are good practical reasons for studying hibernation in a primate species. We humans are genetically more similar to lemurs than to any bear or squirrel. Indeed, we actually share the same genes that are activated when fat-tailed lemurs hibernate. If we could learn how to switch them on, it could have important implications for human medicine. Science fiction is full of stories involving suspended animation for space travel and the like. While this possibility is still a long way off, studies like these are a step towards understanding the mechanisms involved. And maybe along the way, we will discover whether lemurs dream!

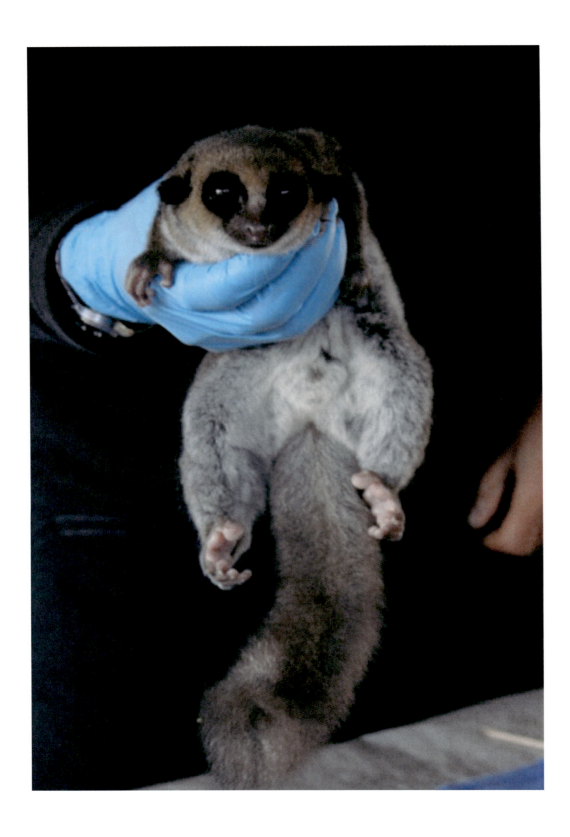

Made in the USA
Middletown, DE
28 January 2025

70203714R00018